AF270697

Twilight Animals

BEARS

by Elizabeth Andrews

Cody Koala
An Imprint of Pop!
popbooksonline.com

abdobooks.com

Published by Pop!, a division of ABDO, PO Box 398166, Minneapolis, Minnesota 55439. Copyright ©2023 by Abdo Consulting Group, Inc. International copyrights reserved in all countries. No part of this book may be reproduced in any form without written permission from the publisher. Cody Koala™ is a trademark and logo of Pop!.

Printed in the United States of America, North Mankato, Minnesota

052022
092022

THIS BOOK CONTAINS RECYCLED MATERIALS

Cover Photo: Shutterstock Images
Interior Photos: Shutterstock Images

Editor: Grace Hansen
Series Designer: Laura Graphenteen

Library of Congress Control Number: 2021951862
Publisher's Cataloging-in-Publication Data
Names: Andrews, Elizabeth, author.
Title: Bears / by Elizabeth Andrews
Description: Minneapolis, Minnesota : Pop, 2023 | Series: Twilight Animals | Includes online resources and index
Identifiers: ISBN 9781098242060 (lib. bdg.) | ISBN 9781098242763 (ebook)
Subjects: LCSH: Bears--Juvenile literature. | Bears--Behavior--Juvenile literature. | Twilight--Juvenile literature. | Nocturnal animals--Juvenile literature. | Nocturnal animals--Behavior--Juvenile literature.
Classification: DDC 591.518--dc23

Cody Koala

Pop open this book and you'll find QR codes like this one, loaded with information, so you can learn even more!

Scan this code* and others like it while you read,

or visit the website below to make this book pop.

popbooksonline.com/bears

*Scanning QR codes requires a web-enabled smart device with a QR code reader app and a camera.

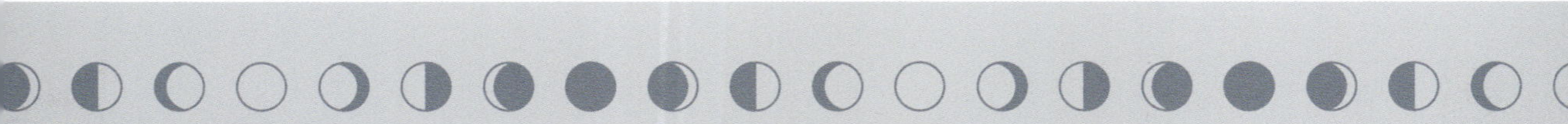

Table of Contents

Chapter 1

Types of Bears

There are eight kinds of bears. They are the North American black bear, brown bear, Asiatic black bear, giant panda, polar bear, sloth bear, Andean bear, and sun bear. They live all over the world.

Panda Bear

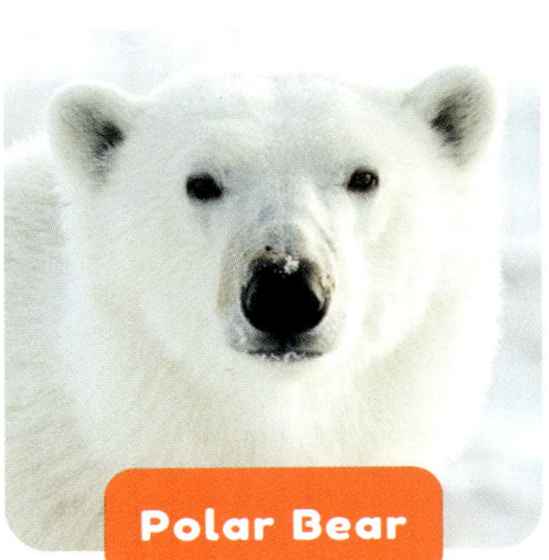

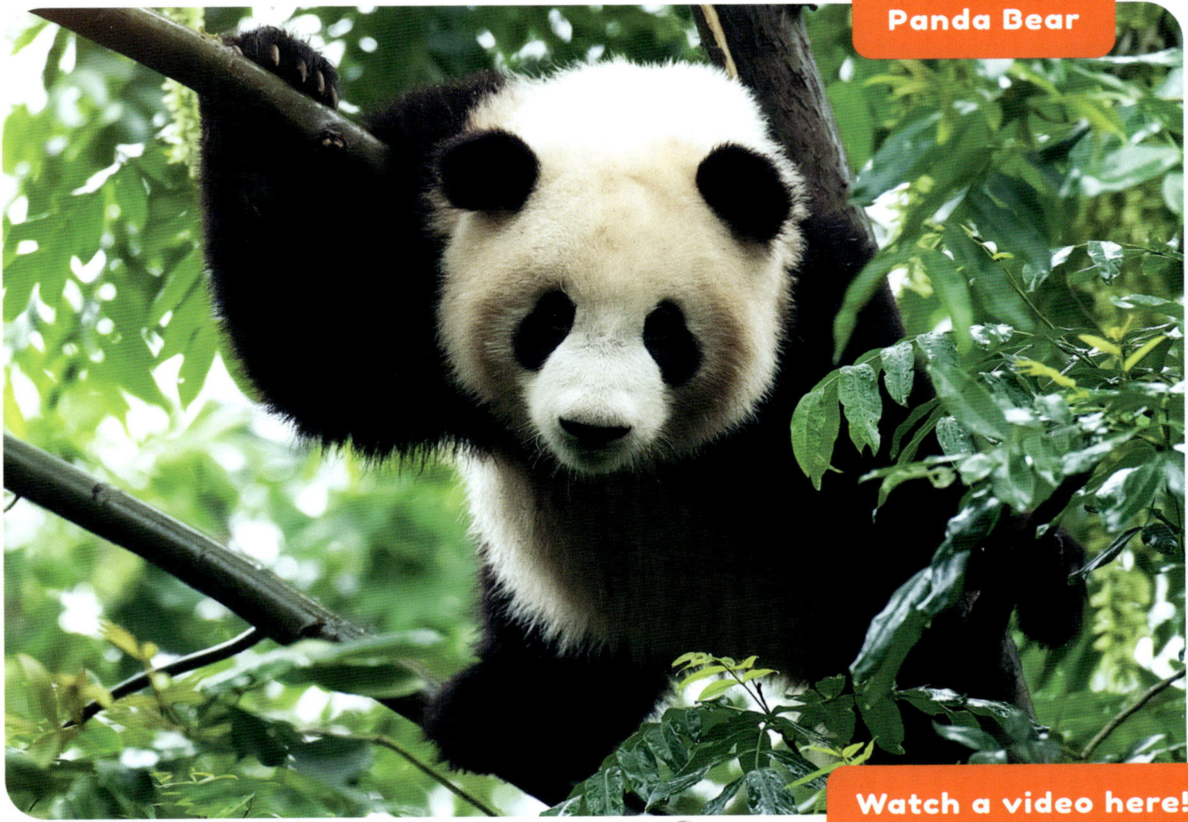

Watch a video here!

Polar Bear

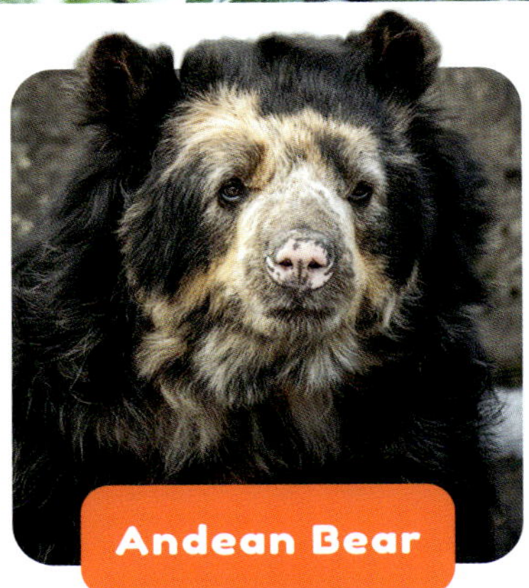

Andean Bear

Chapter 2

Twilight Bears

Brown and black bears are most active at twilight. They don't like the heat of daytime. Black bears **snuffle** through bushes in forests. They look for berries to snack on. Smell is their strongest sense.

During the daytime, bears usually nap in cool, shady spots.
Learn more here!

Black bears are usually smaller than brown bears. They have smaller heads and big ears. Brown bears can get much larger.

Their heads are big with
little round ears. They have a
hump on their shoulders that
black bears do not.

Black bears mostly **inhabit** forests where there are lots of plants to eat. Brown bears live in open spaces. They are found near water because fish are one of their main food sources.

Chapter 3

Sleepy Bears

Black and brown bears **hibernate**. Their body temperature and heart rate drop. Their breathing slows. They can go seven months without eating, drinking, or going to the bathroom.

Explore links here!

When there is a lot of food, brown bears can gain 4 pounds (1.8kg) a day!

Bears must eat a lot during the spring and summer to prepare for hibernation. Bears will eat plants and animals. When they are sleeping, they survive off the layer of fat they built during the spring and summer.

The bears spend the cold winter months in dens they dug or found. If the weather gets warm or something **startles** the bears, they will wake up. They can react quickly and defend themselves if there is a threat.

Chapter 4

Mama Bears

Bears are **solitary** creatures. However, bear cubs will stay with their mother for 18 to 24 months. The cubs are born in the den during **hibernation**. They drink their mother's milk for about two months.

Complete an
activity here!

When the mother and cubs leave the den, the cubs start learning. The mother will teach the cubs how to hunt for food and survive in the wild.

Making Connections

Text-to-Self

There are eight kinds of bears. What bear is your favorite?

Text-to-Text

Have you read any other books about bears? If so, what did those books have in common with this one?

Text-to-World

Bear cubs stay with their mothers for nearly two years. Do you know of any other animals that stay with their mother that long?

Glossary

hibernate – to sleep or rest during cold winter months.

inhabit – to live in a place.

snuffle – to sniff loudly and repeatedly.

solitary – living alone.

startle – to be frightened or surprised suddenly.

Index

Online Resources

popbooksonline.com

Thanks for reading this Cody Koala book!

Scan this code* and others like it in this book, or visit the website below to make this book pop!

popbooksonline.com/bears

*Scanning QR codes requires a web-enabled smart device with a QR code reader app and a camera.